The Hoverboard DIY Guide

Build your own hoverboard

Table of contents

Getting started

But before we get going, and dive into the adventures of hoverboard, let's set the stage. Did you ever chance upon, while surfing on the internet, *Mattel's Back-to-the-future-styled* hoverboard's prop replica selling on amazon.com or some toys merchandising website? Of course, this one doesn't hover. But crealev.com features a new floating hoverboard for sale which comes with their patented *magnetic levitation technology*. Did you see that? However, I bet you saw the video tutorial of hoverboard made by ZA® (Zapata Racing) which is on-display for pre-order on aquaticaviation.net, priced at $5,800.00. Didn't you? Well, I thought you did.

The market, if you explore, is replete with myriad of such products. Toy companies are investing in R&D to develop a hoverboard which can actually hover. May be, the next big toy of the next smart generation will be the hoverboards, replacing the cool skates of our time.

This eBook is your portal into the idea hovering over a hoverboard and a DIY (Do-It-Yourself) manual to allow you to create your own hoverboard using some raw materials. As you go through the book, you will grasp the history and hype hovering around the hoverboards. You will also learn the mechanics behind such hoverboards as you go on to create one of your own with the craft of your own hands.

I just want to mention that this DIY manual will NOT teach you how to hover but how to make a hoverboard. If you're a sci-fi fanatic, easily surrender to the spell of technology and, no matter how hard you try, cannot resist to the temptation of adventure, this eBook is definitely for you.

The science of hovering

Hovercrafts have been in existence for more than half a century now, assisting coastguards, facilitating transportation and, yes, exhilarating in Formula 1 of hovercraft racing. There is no doubt in the existence of hovercrafts. Not at all.

In 1990s, the man named *Robert Zemeckis*, the director of *Back to the Future II*, popularized a concept that was inherently so cool that no one would want to deny it. He declared hoverboards to be real. He fuelled the rumours by saying that hoverboard was in fact real, but was not being marketed because it was seen as too dangerous by the parents. Of course, it was a farce, and it did break the hearts of many children and their fantasy of a hovering on a hoverboard.

Hoverboards were by no means real, not at that time. However, the event did not dishearten some people; they were the people who saw an opportunity – an opportunity to create something new. It was time to go back to the very nature of hovering in hovercrafts, get some inspiration and emulate it for hoverboards.

Hovercrafts: Design principle

It isn't an easy task to get the whole idea of a complex thing like hovercraft fit into a few words, but I guarantee you will understand the basic mechanism which we will employ in building a hoverboard in the part 4 of this eBook.

The idea which allows a hovercraft to hover is *air cushion*. It is for this reason that hovercrafts are also referred to as Air Cushion Vehicle (ACV). If you are lifting an object a little from the ground using air, you are reducing the area of contact between the object and the ground, which causes the friction to reduce in proportion. Since friction is reduced, it becomes easier to propel and move the object. To enhance the capability, the air cushion was made up of a flexible skirt so that small rocks do not slow down or damage hovercrafts. It is this simple principle which helps a hovercraft to traverse hard and rough terrains and even hover over a water surface.

What parts are there?

From the perspective of making a reliable and well-to-do hoverboard, I would mention two basic parts of a hovercraft.

1. **The hull**: It is the main structure of the hovercraft around which all other parts of hovercraft are erected. It has to robust, light-weight and reliable for proper hovering effect. It is below this part that the skirt system is installed.

2. **The lift system**: It consists of an air blower, which feeds into air continuously into the plenum chamber and creates the air cushion necessary to generate the lift for proper functioning.

One last thing which needs consideration is the *power-to-weight* ratio. Although it might not seem critical, it certainly affects the lift system of your hovercraft, and it certainly will be a major factor when you will be designing your own hoverboard.

The power generated by the air cushion must be large enough to produce the levitation in the hovercraft. So at the design stage itself, power-to-weight ratio must be taken in to account to not make things difficult for you. The better you produce the levitation, the lesser friction will be tackled and, hence, better propulsion of the hovercraft. However, care must also be taken so that the levitation is not very high which might make the whole structure unstable.

Now that you are almost educated in the science of hovering, it's time to make a hoverboard, on your own.

DIY-kit for your hoverboard

Before we embark on this journey to develop a hoverboard, just make sure you have the following materials in or around your workplace or at least in your vicinity. You should have all the material available, so that you don't lose enthusiasm of developing the hoverboard by avoiding wasting time in procuring the materials while you are engaged in your work.

1. **Lift magnets**: You need four, 90 pounds each, magnets of circular shape which will play an important part in the design by creating electromagnetic fields necessary for producing levitation effect.

2. **Battery**: You must have two 18 volt rechargeable battery-packs with USB built into it.

3. **Wires**: Considerable length of wires to make suitable connections between magnets, battery pack and the toggle switch.

4. **Toggle switch**: Toggle switch is no different than a natural home switch with button except that it comes with a mechanical lever to toggle between the ON and OFF state. Since toggle switch comes in a variety of shapes and sizes targeting numerous applications, make sure you select a one which can be feasibly attached to the hoverboard without causing any hindrance.

5. **Plywood board**: Cut a plywood board for you with a dimension of 12 x 28 inches with smoothened surface so that magnets can be easily glued to the surface.

6. **Metal Board**: Find a metal board with a shape and size exactly similar to that of plywood board. Select a metal board which is light weight, preferably under 12 pounds, and conductive. It is very crucial that you decide on a metal board of appropriate weight, because it will affect the power-to-weight ratio.

7. **Tarpaulin cushion**: Tarp is developed from a variety of materials. For example, polyethylene, canvas, vinyl, silnylon etc. Try selecting a tarpaulin like polyethylene which is 100% waterproof. But going with canvas type wouldn't do much harm.

8. **Clamps**: Though it may not seem in the initial stages, but it is clamp which will give you the stability to stand sturdy with balance on the hoverboard.

9. **Solder**: Keep a solder to make the connections between wires, magnets, battery and toggle switch. Although it is recommendable to have flux and soldering iron on your worktable, you can still make a good joint with it with skilled hands.

10. **Glue gun**: Keep a glue gun, to glue the magnets with the boards. There are many other ways in which it will be an important resource of your DIY kit.

11. **Re-chargeable leaf blower**: A simple leaf blower will not cost you much, and try using eBay to find a cheap one. Although having a rechargeable leaf blower is preferable, you can use an electric blower and, of course, with only a drawback that your area of hovering will be as wide as the length of the blower wire.

12. **Drilling machine**: To drill holes into the plywood board, you need a drill machine. The machine need not be of large cutting radius.

13. **Clamps**: Choose clamps in which your foot can fit in easily. Not too tight, not too loose.

14. **Duct tape**: It will be a useful asset at many junctures during the design, especially when you need to fill in the openings.

Developing the hoverboard

Design of the hoverboard

The whole structure of the design is differentiated into three layers. This setup offers two benefits: one being that you will be able easily able to follow the design and two, that it becomes easy to develop the design itself. The three levels are explained below

Level 1

The very first and top layer of the hoverboard comprises of a mini leaf blower with an 18 volt rechargeable battery pack along with a built-in USB and a toggle switch soldered to the battery pack. It also has a hole through which the leaf blower blows. The shoe clamp is also fixed on the top layer so that you can stand stably on your hoverboard.

Level 2

The second layer contains no more than two pair of magnets which are connected with the toggle switch through wires. The magnets are placed at a certain position as you shall see, and face opposite to each other. Around the wooden board is a circular tarpaulin fabric which is inflated using a blower

Level 3

The third and bottom layer is made of a metal board with size same as that of plywood board and attached with two small holes in it.

Designing the hoverboard

This section contains a sequential step-by-step method to build a hoverboard using the DIY-kit mentioned in the previous chapter.

Step 1: Setting up the board

Cut down the board to an exact size of 12 x 28 inch. I chose to rely on this size because it offers better stability and smoother control. You can cut any shape you like, but for simplicity and avoiding complications, I have chosen to use a rectangular shape.

Now, I assume that the board has sharp edges. There is no problem in having sharp edges; however, I would like you to add some curve at the edges of your board. The purpose is that it doesn't end up damaging the tarpaulin sheet with which you will cover it. Try opting for a radius of 8-10 cm which will also allow you to wrap the edges easily with a tarpaulin sheet as we will see in a few steps from now.

Advice: Your board is meant to lift your weight, so it must be evident that you need a strengthened board. If you only have thin plywood, you can cut two ply boards of 12 x 28 inch (make sure that the two sizes are similar with high preciseness) and screw them together.

Step 2: Drilling a hole into the plywood board

There are many ways and tutorials on YouTube where you can find myriad of ways to cut a hole into a wooden board, but I prefer a drill for it is easier to find and operate. You can use a hole-saw if you want. Since we wanted to fit the blower pipe into this hole, the hole in the board must be comparable in size to that of the opening size of the blower. The hole must only be large enough to just fit in the pipe of the blower.

Advice: Drilling sometimes can get tricky, especially for the novices. Here are some precautions to take while operating a drilling machine.

Step 3: Gluing the magnets

It's time now to use your glue gun! No, No. Not so fast. You will be soon gluing two pair of magnets to the plywood board you just cut out. But before you take that gun into your hands, mark the centre of the board along the 12 inch breadth of the board by drawing a straight line with a pencil. Also mark two points, each at a distance of 6 inches from the centre (along the length of the plywood board) but on the opposite side, marking the rear and the tip end of the board as A & B, respectively.

Now you can take that glue gun, and add considerable amount of glue to the rear end (point A) of the plywood board and fix a pair of magnets (90 pound each) just ensuring that the centre of the circular magnet lies on the line you just drew.

Advice: Make very sure that these magnets, fixed at the rear end of the board, are glued well because it is from here that the most of the power will be generated for the lift.

Now do the same with the tip of the plywood board and glue another pair of magnets (point B) as well while keeping in mind the same precautions as above.

Step 4: Making the circuit

Cut a pair of wires of equal length. Now with the solder and using a soldering wire, solder one of the wires to the positive terminal of the 18 volt rechargeable battery and other to the negative terminal.

Solder the ends of these wires to the positive and negative terminals of the toggle switch respectively. Now you need two more wires through which you will connect the toggle switch with the two pair of magnets on the opposite end of the plywood board.

Now, using the glue stick, attach this whole set up of wires, battery pack and toggle switch to the hoverboard. If you are thinking of placing it on the side of the board, reject the idea because it will be covered by the tarpaulin wrapping.

Advice: Soldering can get tricky for some of few, especially if you never have had hands-on experience. If you have never done soldering before or still carry a hesitation in handling solder and soldering wire, here is a tutorial for you.

Step 5: Setting up the metal board

Before you go to fix metal board with the system, make 2 holes at the bottom of the metal board to allow the air to escape from the air packet. You can either use a duct tape or a file to smoothen the edges of the metal board. Now attach this metal board the bottom of the structure already made by screwing the metal board properly in multiple places.

The metal board must be attached in such a way that it gives equal pressures at all points of the tarp cushion.

Advice: The metal board you chose for your hoverboard must not be very heavy which may risk the very levitation upon which the hovering depends.

Step 6: Wrapping in the tarp sheet

Most of the work is completed now. Now spread the tarpaulin sheet on the floor and place the three layers of the design together on it in such a way that the tarp sheet is evenly spaced at all sides.

Before wrapping the cushion, cut some holes through it for allowing the air to exit. The air exiting from the wrapping cushion must also have an even spread.

Now fold the tarpaulin sheet over the board and fix it firmly with the level 1, and make sure that there is no opening left in any way, for it may result in uneven leaking of air from level 1 resulting in faulty operation. Moreover, you wouldn't risk the tarp sheet tightened on the base of the metal board. An inch or one-and-a-half inch would be good for your hoverboard.

Advice: The amount of air exiting through the holes must not be lesser than the amount of air entering through the hole at the level 1 by the action of blower. For if it happens, you may risk the tearing of these holes altogether.

Step 7: Fix the blower

Now take the blower and fit it into the hole you made onto the plywood board. To make sure that there is no opening and in case the hole outsize the blower opening, you can use the duct tape to fill. Next to it place your leaf blower depending on the type you are using. It is very important to choose the blower position carefully for it affects the stability of hoverboard while it hovers.

Step 8: Bind the clamp

It is almost done now. Take that clamp you have and fix it to the wooden board using screws. Be careful before you place that clamp. It must be in a suitable position for you to fit that foot in.

Advice: Before you attach it with the top layer, try standing on the *almost-complete* hoverboard while trying and checking the most suitable position to fix the clamp.

Step 9: What now!!

Nothing! I was just testing you patience. Get on that hoverboard.

And fly.

Advice: Choose an open space for your first endeavour. Try keeping your weight as central as possible and the centre of mass as low as possible.

Improvements and future

I hope you went all the way through with me to design a hoverboard. Probably, you already have had a ride or two on it, and planning to show it to your friends and, there is a good chance that you might even flaunt it; and why not? You did it all by yourself, worked hard, put in your mental efforts and made science proud.

But that's not all. You can still do a lot with the design, if you want and take it to another level. Here are the two things:

Aesthetics

To make it simpler for those who are not good with paint, I refrained from adding some paint work to this hovering board. You can add some artistic look to your hoverboard and make it look even better and fancier. There are many designs available on the internet or you can use the vastness of your own imagination.

The leaf-blower

If your leaf blower did not generate enough lift to create a hovering effect, you can use a more powerful leaf blower enhance the hovering experience. You can also do so by choosing a still lighter metal board which denotes the level 3 in our design.

Future of hoverboards

There is one final word left to what you can do. Hendo, a start-up of two couples, has developed a hoverboard which promises everything which *Back to the Future II* showed us two decades ago.

You can see the product demo and Kickstarter for the hoverboard campaign here. You can get the hoverboard's primary product, Whitebox Developer Kit, at $300, which includes hover engine along with some additional materials so that you can experiment with the board yourself.

The campaign is still backed by more than 25,000 people. The Hendo allows its buyers to put some effort into the design to make it a reality, rather than develop the product themselves.

There is a lot to be done to make hoverboard a commercial reality. No matter, what efforts it takes, the hoverboard will materialize soon. And the best thing is that *you take part into making hoverboards a reality*.

The Hoverboard DIY Guide

Build your own hoverboard

Table of Contents

Foreword

Man has no limits. While most phrases has become a cliché, such that their pronunciation barely evokes any emotion, "*man has no limits*" is seems new every time one enunciate it. We are breaking boundaries, reaching new horizons, exploring arenas which were locked since the existence of the planet. New avenues are sprawling up. The world is changing. But one thing hasn't: *man has no limits*.

While I was embarking on this eBook, I had this in my mind. My sole purpose of writing this eBook is to give impetus to this thought. You have all that you need at your disposal. You can use them to create things that humanity might not have imagined.

In this eBook, I have tried to equip readers—the general readers, and not just the scientific readers—with the knowledge and gumption to build their own hoverboard.

With the new revolution in the arena of transportation, with Uber and Flyt taking over the public transportation, the transportation is going to change in the years to come. Hoverboard is one such possibility. We already know about Hendo hoverboard and the recently launched Lexus hoverboard (it actually came into the news while I was writing this eBook). While the product cannot be marketed at this time due to high cost and inherent problem with the product, (it works only on special surfaces) the beagle has been sounded. And don't you think it's a baby step. It is a leap.

When *Back to the Future* envisioned hoverboard, everybody wanted it to be a reality; the kids even more. Many hoverboard toys were launched as well. Some went on to say that the hoverboard is impractical; that it cannot be materialized. I guess, whosoever said that must be no different than the guys who said that we cannot reach the moon. Or the guys who maintained that we cannot talk to people at a distant land. Or those who asserted that human cannot fly.

I guess they could understand that *man has no limits*. That Pluto is not far away. That we have already reached Mars. That quantum computers are possible. That artificial intelligence will be our greatest creation ever. That everything is possible.

And so is our hoverboard.

I hope you are ready and excited to read this eBook. It would be a great journey ahead as we explore new horizons and build our own hoverboard.

Introduction

When ArxPax came up with Hendo hoverboard, powered by ArxPax Magnetic Field Architecture (MFA™) technology, they did make a mark on the minds of the people. Why, ArxPax raised over $510,590 in the last December on Kickstarter, a crowd-funding platform.

The Hendo hoverboard floats about one inch above the ground using electromagnets. Photograph: PR

Even though ArxPax has some ambitious goals like levitating the buildings, homes, and schools to escape natural calamities, the hoverboard project is improving. They are now trying to take it next levels.

With an aim to make the hoverboards human-driven, Hendo is now working on to find new compounds. The purpose is to make the hoverboard technologically better and minimize the cost at the same time.

The Lexus hoverboard
It is all in the news. The car company, Lexus, has recently launched its first hoverboard. The video they launched showing their hoverboard will make your

eyes still. It looks finished, flawless, and fabulous. Unlike the ArxPaxHendo hoverboard, this one can actually be ridden.

The speculations tells that the hovering effect is of Lexus hoverboard comes from the magnetic field trapped in the semiconductor.

Lexus hoverboard

What will we study in this eBook?
The Hendo hoverboard as well as the Lexus hoverboard need special surface to get the levitation.

Both hoverboards may look similar, but the design and the technology behind the two is different. While the use of magnetic field is obvious in the two, the science is different.

The Hendo hoverboard uses electromagnetic repulsion to generate the levitation. On the other hand, Lexus hoverboard uses a combination of superconductors and Meissner effect.

We will study the science behind the two hoverboards and try to unravel how these technologies work. We will also discuss a few other technologies which can be deployed to make the hoverboard possible.

We will start by discussing how levitation, which is crucial to hovering, works through various scientific explanations.

We will then make our way to create our own hoverboard using the technologies we discussed. Don't worry, the material will not be as costly as the superconductors. You will be able to avail the required tools and materials to build the hoverboard.

Who should read this eBook?

I hope I didn't scare you with semiconductors, Meissner effect, and other science jargons because I am writing this eBook so that everyone can read and know the fascination involved in hovering.

If you think you will be bombarded with jargons and science that may go above your head, wait till you go ahead. We will discuss everything in simplest terms so that anyone can follow it easily and build his own hoverboard.

The only requirement to read this eBook is your curiosity. I hope you have it in abundance.

The Science of Hovering

So, the big question: how can we make a board hover?

Even though hovering is not a rocket science, it still is a wide field whose doors have not been unlocked yet. Consequentially, it is almost like a rocket science.

To make a board hover, you need to do two things;

1. Lift the board, and
2. Propel it!

It sounds simple, doesn't it? I leave it on you to decide.

We can divide the hovering science into two other sciences—one of which deals with lift while the other with propulsion.

In this chapter, we will discuss various sciences of levitation and propulsion: Magnetic levitation, quantum levitation, ionic-airlift propulsion, Heim theory, electromagnetic resistance, and super conductivity. We will see how these theories can bring about levitation and propulsion effect, which in turn will get our hoverboard going.

Magnetic and paramagnetic levitation

If you have completed your graduation in physics, there are good chances that you are aware of magnetic levitation. Does maglev rings a bell? Maglev is simply the annotation for *magnetic levitation*.

Also known as magnetic suspension, maglev is renowned method of suspending an object in the air with the help of magnetic fields. There are practically no other forces involved. The idea is to overcome the gravitational forces (or acceleration) which is responsible for keeping objects on the ground. Maglev is already in use and has been deployed in super-high speed trains.

How does magnetic levitation works?

Well, to make maglev work, there are two necessary requirements.

1. Lifting Force: A sufficient force to keep an object suspended by countering the gravity
2. Stability: Keep the object stable by avoiding slide or flip, which would eventually neutralize the lift effect

Transrapid series 09 vehicle at the Emsland Test Facility, northern Germany. A shining example of maglev. It can go as fast as 500 km/h!

Lift

A 1st standard kid will be able to tell you a few things about magnet; like two magnets can either repel or attract each other, based on their orientation towards each other.

The same holds for other magnetic materials and magnetic systems. They are able to attract or repel with a force that is proportional to the magnetic field. The area of the magnet or the magnetic system also affects the force.

A lift effect may easily be created by using two dipole magnets positioned in such a way that their like poles face each other. The force is often quantized as magnetic pressure. Mathematically, we define magnetic pressure as:

$$Pmag = \frac{B^2}{2\mu}$$

$Pmag$ denotes pressure in Pascal, B denotes the magnetic field in Tesla, and μ denotes permeability in the vacuum, which is a constant equals to $4\pi \times 10-7$ N·A−2.

The greater is the magnetic field, the greater will be the magnetic levitation. But how can we generate high magnitude magnetic fields?

I think I will let the suspense suspend for a while...until we start designing out own hoverboard.

Stability

Even if you have gained some lift, it is of no use if you can't stabilize it. The simplest lift we discussed above is highly unstable. The top magnet will easily slide, slip, or flip destabilizing the whole configuration.

However, the researchers have been able to achieve stability in the lift by using servomechanism, diamagnetic materials, super conduction, and even eddy currents.

Paramagnetic levitation

There's one more facet to maglev—paramagnetic levitation. Paramagnetism is a property by virtue of which certain materials are attracted towards a magnetic field and induce internal magnetic fields which are directionally towards the applied magnetic field.

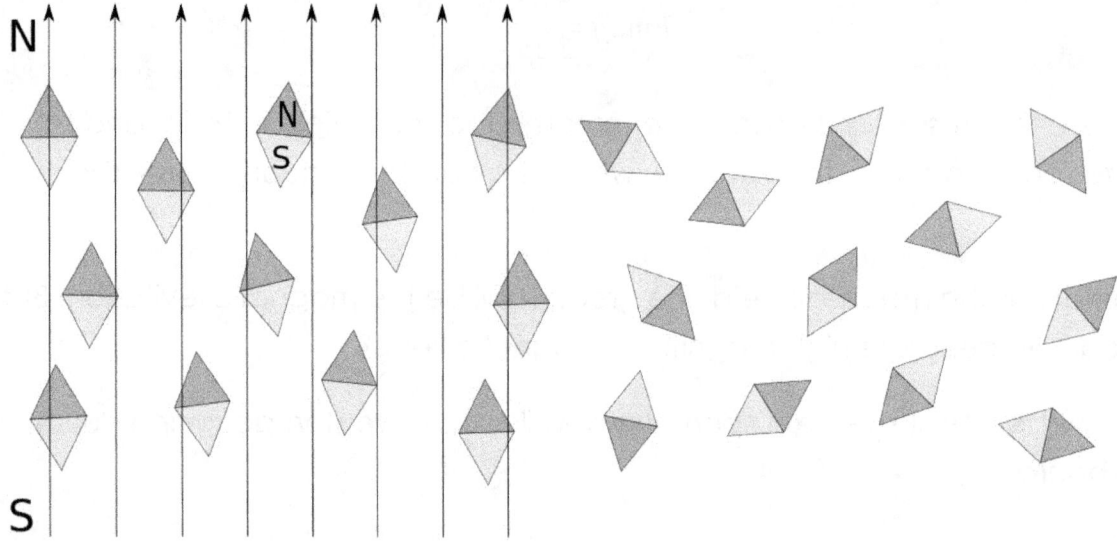

If provided right conditions, Oxygen (O_2) can show paramagnetism. The idea is if a gas can be magnetized in this way, it can be contained. All that would be required is to supply gas continuously to compensate for what leaks out.

The only problem is: paramagnetism is weak. As thermal motion randomize the gas particles, a marginal fraction of molecules are aligned.

Quantum levitation

Quantum levitation comes from one of the classic discoveries of science—quantum mechanics. It involves using the principle of quantum physics to give lift to an object.

The quantum levitation theory we are about to discuss now demands an overview of Meissner Effect.

Meissner Effect

Walter Meissnerand Robert Ochsenfeld discovered a phenomenon in 1933 during an experiment. They were trying to measure magnetic field around various materials. They found that when some materials were cooled to superconductors, the magnetic field was almost negligible. That is, superconductors would negate all the magnetic field inside it.

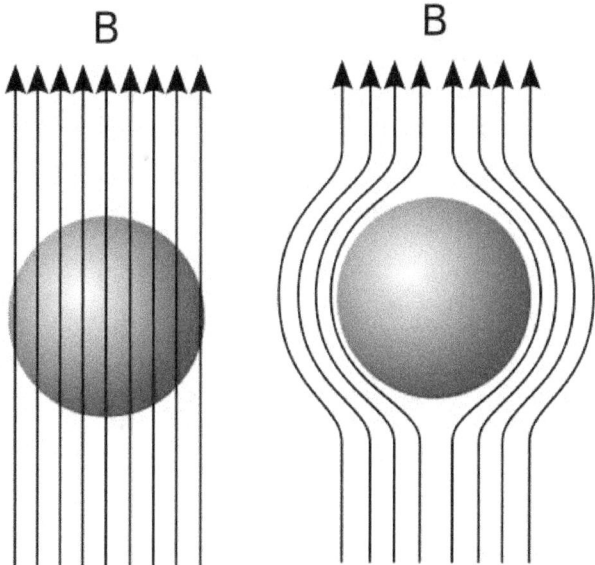

Meissner Effect: *B* in the above diagram represents magnetic field. **The left side shows a normal conductor, while the one on the right represents a superconducting material.**

This negation of magnetic field inside of superconductors is due to the small currents induced at the surface of superconductors in the presence of a magnetic field. The current is easily produced due to absence of no resistance. The current, however, produces a magnetic field which cancels out the magnetic field inside.

This results in a bend of magnetic field around the superconductor, creating a wrapping effect around it.

How does it cause quantum levitation?

Now, suppose you place a superconducting material above a magnetic track. According to the theory we discussed above, the superconductor should be pushed away by strong magnetic fields. Hence, it would stay slightly above the track, as much as the opposite gravitation allows it to be.

Yes…I know it sounds simple. But it is not over yet. We still have a problem of stability. As in paramagnetic levitation, quantum levitation will also make the superconductor slip or slide away from the magnetic field. The levitating object will not hold its place. It's similar to what happens when you place two magnets with their south poles or north poles facing each other.

However, there's a solution. Enters the scene…quantum locking.

Quantum Locking

There are two types of superconductors: type I and type II. Only type I superconductor exhibit pure Meissner effect. On the other hand, type II superconductor doesn't display diamagnetism perfectly.

"Meissner effect p1390048" by Mai-Linh Doan - self photo.

In type II superconductors, some of the external magnetic field penetrates the superconductors. This penetration is actually the key to quantum locking.

All those spots or areas where magnetic field penetrate the superconductors are called "vortex". Apart from type II superconductors, flux vortex can also form in thin superconductors made out of type I superconductors. The flux vortex effect can be enhanced with type II semiconductors.

So, how does flux vortices lead to quantum locking and eventually stabilization?

The areas where flux vortices form or where magnetic field penetrates, the superconducting is turned off in the narrow region. It's like a superconductor with lots of non-superconducting region. With movement of the superconductor, flux vortices also moves.

Because of the Meissner effect, the superconductor itself will exert a force inhibit any motion that cause change in the magnetic field. Any change in the position of superconductor will cause it to return to the same spot.

And that's quantum locking! The lifted object will not wobble any more.

Ionic airlift propulsion

Let's talk about another thing that hovers: ionocraft, also known as ion-propelled aircraft. The best part is that it requires no combustion for its operation.

So, how does ionic airlift propulsion works and why it is important for our discussion?

It is important for two reasons. Firstly, that it hovers; secondly, that it moves. The last two systems that we discussed talked only about the lift. They were totally silent about the movement. How will we make a hovering board move?

Ionocraft has answers to most of our questions.

How does it work?

It is based on a propulsion mechanism called ionic air propulsion. It was Francis Hauksbee who first came up with the idea of ionic air propulsion way back in 1709. It was mentioned in his book *Physico-Mechanical Experimentson Various Subjects.*

Heim Theory

Burkhard Heim can be credited to have taken one of the most challenging tasks in physics. In 1957, Heim tried to produce a *theory of everything.* He tried to resolve the two most spectacular theories in theoretical physics: general relativity and quantum theory.

He produced a mathematical approach to quantize space-time which will eventually lead to reconciling general relativity and quantum theory. Einstein came up with 4-dimensional framework to equate matter with energy; mass can be converted to energy and vice versa. Heim wanted to propound a theory that would convert between all kinds of energy. He added two extra dimensions in Einstein's framework, and were later expanded to include eight and twelve dimensions.

However, Burkhard didn't get much attention. His theories have never been in the mainstream for long and is as close as dead.

But not everything that Heim had said was inconsequential. There are ways in which it can be used to materialize new possibilities. Heim had some success in this attempt.

What does Heim Theory mean for the future of hovering?

Heim had succeeded in establishing one thing, at least in theory. He was able to equate electromagnetism and gravitational energy.

If you have a knack for sciences, you can easily understand what it means for us who are trying to build a hoverboard.

If you can equate two parameters, it simply means one can affect the other. So, according to Heim Theory, electromagnetism and gravitation energy can have effect on each other.

We are constantly under the pull of gravitation. It is the one thing that keeps us bound to the surface of the earth. If we can overcome this pull and neutralize it effect, in some way, we can certainly get a lift. Heim also had this notion that if we can generate enough magnetic field, we could easily break free from the grip of gravity.

How can Heim Theory be used to make an object hover?

Walter Dröscher is one of those rare scientists who took interest in Heim's theory and collaborated with Burkhard Heim to develop it further.

Walter Dröscher worked along with Jochem Hauser to create a theory that aimed to power the spaceships; and which we can use to make hoverboard work.

The idea goes like this: Take a big rotating ring and place it above a superconducting coil. According to them, if the ring is rotated and a high current is passed through the ring, the magnetic field generated will cancel out the effect of gravity on the ring. And as you can guess, the ring will fly or hover freely.

Simple. Put still not practical.

The bad news is that we still lack the technological development to create such a current density. Unfortunately, the theory remains in fog without being tested; at least for now.

Electromagnetic repulsion

You might have learnt about it during your high school, or probably as your hobby. It is such a revolutionary and "plausible" concept we have. We can also use it to get our board hovering.

How can we use it to build a hoverboard?

The idea is extremely simple: opposite charges attract; like charges repel! It can't be any simpler.

Starts by creating an electrostatic force between two objects, which in turn depends upon creating electrostatic charge. Once the charge is high enough, the two objects will repel.

In our case, using a metal surface, we can easily design a hoverboard which can hold high charge; and to be precise, positive charge, because the earth's surface carries a small amount of positive charge.

The force of repulsion will depend on the proximity and the charge. It is governed by the following equation.

$$F = k \frac{Q1Q2}{d^2}$$

F denotes the force. k is known as Coulomb's constant which equals (8.988 x 109 N.m²/C². Q1 and Q2 denotes the charge on the two objects while d denotes the distance between them.

Like charges repel: This image shows how two positively charged object will repel each other

The problem

The only problem is that the earth's surface or the ground can hold only a very small amount of positive charge. We can overcome this problem simply by increasing the charge of the metal surface, but that is not enough as well.

If we do try to increase the charge to a high value, the problem is still not resolved. See, the hoverboard is extremely near to the earth's surface. The high charge induces an opposite charge to the nearby object. In our case, it will cause the earth surface in its proximity to be negatively charged.

That means you will have to face an attraction you don't want. The hoverboard will not hover at all.

Moreover, the proximity of the hoverboard with the earth can also cause the charge to be discharged into the ground.

What to do now?

Well...you can do a lot. We have science watching our back. It is going to give us a lot of support to make our dream of a hoverboard come true.

In the next chapters to come, we will use the theories we discussed to build our own hoverboard.

For now, I just want you to recall and keep in mind all the theories we discussed above. Apart from designing this hoverboard, you can build on them to bring innovation and creativity into the hoverboard. Whether it is electrostatic repulsion or no-so-much-appreciated Heim theory, you need to keep these concepts in mind before we embark on the next step.

What Do We Need?

In this chapter, we are going to look at the requirements—what we actually need, apart from science, to make the board hover.

You might think we should discuss the design before we go on to discuss the requirements. I take a rather different approach. The better you understand the requirements, the better you will be able to understand the design and consequently, execute it better.

The ingredients this time will be a little different than the last time. Since, we have improved the design and the technology behind it, we will need a few more equipment. The basics will still be the same this time.

Metal board and wooden ply board

We need a strong and sturdy metal board. But it does not need to be heavy. Because if it is, the power-to-weight ratio will decrease, making the system's power weaker to make a lift.

Similarly, we need a wooden board. The size and weight criteria would follow the rules we have set for the metal board.

Repulsion coil

First of all, we need repulsion coil. And not just one, but two. I hope you can easily find them at an electric store. Otherwise, there are some people who are selling it online.

You can also buy one at JerrysElectronics.com. The only problem is that it has already been made for a specific purpose. You would have to do a little engineering with it to get it ready for our purposes.

Once you have the design, you will know how to use it.

Electromagnets

Then we need two electromagnets as well. Unlike in the traditional magnets, the magnetic fields in electromagnets are produced by the electric current.

It consists of a metal with a wire densely wound around it. The coil behaves as a magnet as soon as some current is passed through it.

30 V battery pack

Now that you know we need current, we need a power source as well. A 30 V battery back would be perfect for our purposes. You can easily buy one from your nearest store. Or there are plenty of options available on Ebay.

Rechargeable leaf blower

Those who haven't read my previous eBook on hoverboard will surely wonder and, may be question, the need of a leaf blower.

Who would have thought that a gardening tool would be so useful in our endeavor? Anyhow, make sure you find a leaf blower, and that too rechargeable. You can easily find a lot of them on the internet.

Tarp lift system

The tarp will also be used in our hoverboard. Pretty much similar to what we did the last time. However, we will also use a tarp lift system this time.

Toggle switch system

A simple toggle switch we need. The one which will be used to power on and off the hoverboard. You do not need to spend much on one. Any toggle switch will do for our purposes.

Wiring

No electrical system is going to work without it. So, I would recommend that you carry a lot of different types of wires with you. They may also come in handy in various other ways.

Other equipment

Then we will need all those little but important things which bind the system. You will need a soldering iron, flux, and glue gun to make the connections. Also arrange a duct tape to bind the system together. Like wires, duct tape can be of help in all sort of ways.

You might also need drilling machine to cut holes in the wooden board. I also assume that would like to mount on the hoverboard and take a ride, wouldn't you? In that case, you will also need clamps to ensure you do not fall on your very first ride.

The Design

Another big question: how will our hoverboard look? How has it been designed? Will it hover for real? How high would it be from the ground? Does the hoverboard make sound? How long will it hover in the air in one go? Is it reliable? Can I use it to hover my way to the office?

Not all your questions will be answered at this stage; but some of them have very definitive answer. We surely know how our hoverboard will be built. The rest you will know once you have built the design.

Now that we have arranged all the materials for building the hoverboard, we can see how all of them are combined together to make this hoverboard.

The overview

The design is not much different to what we had seen in my previous eBook, except a few changes.

The hoverboard is divided into 4 parts or layers. Each of the layer consist of different parts. One by one we will see each of the layer and how it has to be designed.

Layer 1

The first layer is the wooden layer. This layer consists of a rechargeable leaf blower. Why rechargeable? You will know as we move on.

Apart from the leaf blower, layer 1 will also consist of toggle switch. This will give us the control to turn ON or OFF the hoverboard.

Layer 2

The second layer is the metal layer. The metal should preferably be aluminum. This layer will consist of electronic systems pertaining to the hoverboard.

Layer 3

Layer 3 is also made up of metal. This should be above and attached to the second layer 2.

Layer 4
The final layer is more or less a cushioning system which will help in making the board hover. This layer will, however, be merged with other layers, covering it from all sides.

Building the Hoverboard

Now, you have all the materials involved in the hoverboard. You know the design as well. You know the science behind the hovering and propulsion effects.

So, let's kick-off the project.

Before we proceed, let's have an idea of how we should proceed. As you know, we have four layers in our hoverboard, we will follow a layer by layer design approach to build it. Starting from the first layer, we will make our way to the fourth, while keeping the structural integrity of the hoverboard.

Setting up the first layer
The first layer has to be very simple and elegant. It is like the base of your hoverboard system. All the other layers have to be coherent with the wooden layer.

Also, affix a toggle switch at the bottom of the board. (You will know the reason as we move on) Now, precariously add a hole into the wooden board which will be used to attach the lift system.

Glue the electromagnets
Now that the switch is at its right place, it is time to glue the magnets, electromagnets, to be precise. Glue both the magnets at the either ends of the hoverboard such that they are in the center of the breadth. The symmetry is essential here.

Setting up the metal board
The second layer will consist of electronics system. It will also have a hole aligned with the hole in the wooden board.

Now, attach the two repulsion coils and the two electromagnets to this layer. Since, both these elements are large in size, you need to manage the space at this layer. Make sure the system is spacious and connections can be made reliably.

The 30 V power supply has to be added to this layer as well. The power supply will be responsible for powering the electromagnets and make the repulsion coils work.

Setting up the circuit

Bring the wires into the picture. Connect the 30 V battery with the toggle switch. Now, through the toggle switch, take extend these wires. Connect these wires to the electromagnets we glued to the wooden board in the previous step.

Adding the repulsion coils

Take the two repulsion coils together and glue them to the metal board such that one is at the top of another. These two repulsion coils will carry on the effect of electromagnetic repulsion to make the lift possible.

Wrapping the tarp sheet

It is time now to wrap the tarp sheet around the structure. The best way to do it is to first place the tarp sheet on the floor. Then place the hoverboard structure over it. Make sure the tarp is evenly spaced at all the sides.

Now, take the sides of the tarp sheet and wrap it around the hoverboard. It should be enclosed in such a way that the there is no chance for the air to leak. Because if it happens, your hoverboard is never going to make a lift.

It is also important at the same time to ensure that amount of air entering the hole should be also almost equal to the amount of air leaving it. Because if it does not happen, either the holes will burst or the hoverboard will not hover at all.

Attach the leaf blower

The last step is to attach the rechargeable leaf blower to the wooden board. You can use a drill to do that. You just have to make sure that the hole you make in the board must be comparable to the opening of leaf blower. Plus, it should fit perfectly into the hole. Make sure the leaf blower is affixed properly such that it does not wobble or shake when it is turned ON.

Let's hover

Now, add the clamps to the hoverboard. You can simply attach them to the hoverboard using a screw. Just make sure your foot fit into it and the distance between the two clamps is enough to make you stand comfortably.

Everything is ready now! Test your hoverboard.

Does it work?

I'm sure it does! Get on it and hover!

www.ingramcontent.com/pod-product-compliance
Lightning Source LLC
Chambersburg PA
CBHW080621180526
45168CB00007B/3009